Keep Your Students in Mathematical Shape!

Mathercise™
Classroom Warm-Up Exercises

Book D

For Geometry, Advanced Algebra,
Second or Third-Year High School Math

Michael Serra

KEY CURRICULUM PRESS
Innovators in Mathematics Education

Copyright © 1992 by Michael Serra. All rights reserved.
Published by Key Curriculum Press, 1150 65th Street, Emeryville, California 94608.
e-mail: editorial@keypress.com
http: //www.keypress.com
Cover art by Don Dudley. Graphics by Ann Rothenbuhler.

Printed in the United States of America   12 11 10 9 8 7 6   04   ISBN 1-55953-062-6

What is *Mathercise D?*

Mathercise D is a mathematical fitness program designed to keep students who have had a year of algebra and are taking geometry, advanced algebra, or a second or third-year high school math course in great mathematical shape. It is a series of 50 class starters which strengthen important math skills. You can easily integrate *Mathercise D* into your existing math program because it requires very little additional preparation. Instead of taking up valuable class time, a regular *Mathercise* program will actually give you more time to teach because it encourages your students to come to class on time and immediately get "on task."

The book contains 50 reproducible *Mathercise Book D* masters. You can use the masters to create overhead transparencies or reproduce them as individual student worksheets. Each *Mathercise* is a set of three problems and a space for a fourth. The first problem (Reason) is always an inductive or deductive reasoning problem. The second problem (Solve) is a symbolic thinking problem that reviews one of a number of basic mathematical skills (averages, percent, probability, algebra, proportional thinking, functions, and coordinate geometry). The third problem (Sketch) is a visual thinking problem that requires students to think in two and three dimensions and graph linear, absolute value, quadratic, and exponential functions — many with horizontal and/or vertical shifts.

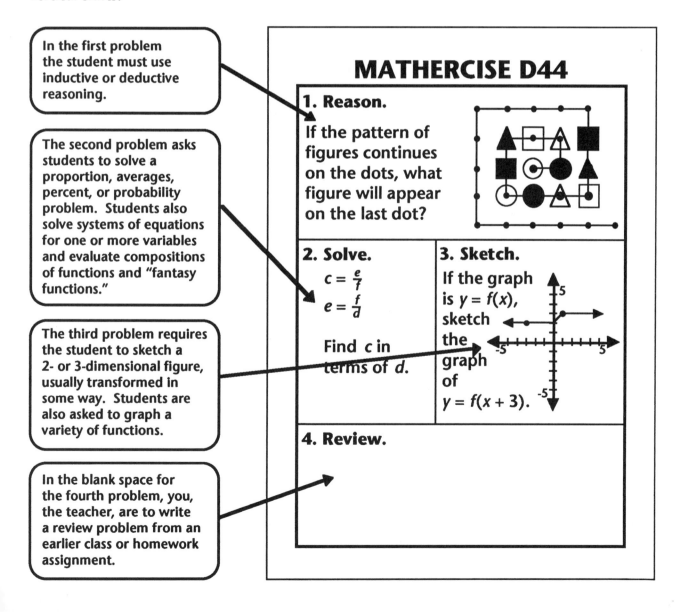

In the first problem the student must use inductive or deductive reasoning.

The second problem asks students to solve a proportion, averages, percent, or probability problem. Students also solve systems of equations for one or more variables and evaluate compositions of functions and "fantasy functions."

The third problem requires the student to sketch a 2- or 3-dimensional figure, usually transformed in some way. Students are also asked to graph a variety of functions.

In the blank space for the fourth problem, you, the teacher, are to write a review problem from an earlier class or homework assignment.

MATHERCISE D44

1. Reason.

If the pattern of figures continues on the dots, what figure will appear on the last dot?

2. Solve.

$c = \frac{e}{7}$

$e = \frac{f}{d}$

Find c in terms of d.

3. Sketch.

If the graph is $y = f(x)$, sketch the graph of $y = f(x + 3)$.

4. Review.

What are the objectives of *Mathercise D*?

• *Mathercise* **gets students in the habit of being "on task" right from the start of class.** All too often the first 5 to 10 minutes of class time is taken up with routine clerical activities. Students learn that it is smart to be tardy. *Mathercise* can encourage more productive study habits and promote punctuality.

• *Mathercise* **gives students the opportunity to develop and practice both inductive and deductive reasoning skills.** Math is supposed to be the subject where students develop reasoning skills. However, we rarely give students an opportunity to build and practice these skills before demanding that they apply them to new topics. *Mathercise* lets students develop reasoning skills in a familiar context, without the added burden of learning new mathematical content.

• *Mathercise Book D* **gives students needed practice with the basic skills of proportional thinking, probability, and the graphing of linear, absolute value, and quadratic functions.**

• *Mathercise Book D* **gives students practice drawing 2- and 3-dimensional geometric figures.** Many students have difficulty drawing basic geometric shapes. This useful skill can be learned with practice.

• *Mathercise* **provides you, the teacher, with an alternate means of motivating students to do their homework and a quick method of checking on student progress.** With busy schedules and large classes it is nearly impossible to look over all the homework of all of our students. *Mathercise* can be used as an alternative or additional check on student progress. By placing an assigned homework problem in the review slot (problem 4) of a *Mathercise*, you can encourage students to do their homework. If students are permitted to use their notebooks for the *Mathercise*, it even prompts them to keep accurate notes.

• *Mathercise* **prepares students for college entrance exams.** The second problem on each *Mathercise* is similar to one type of problem appearing on standardized tests. You can use the review problem to practice additional problem types found on standardized tests.

• *Mathercise* **once a week will keep your students in great mathematical shape!**

How do you use *Mathercise Book D* in your classroom?

Use *Mathercise* as a class starter, getting students "on task" immediately at the start of the class period. *Mathercise* is designed to be used once or twice a week with students who have taken a year of algebra and are taking geometry, advanced algebra, or a second or third-year high school math course. Here are some suggestions that will help you use *Mathercise Book D* effectively.

• Make transparencies from the masters for use with an overhead projector. If you can't use an overhead projector, make individual copies for each student in the class. Students shouldn't write on the copies, but instead should put their answers on a separate sheet of paper. For the third problem (Sketch) require students to sketch the figure at a different size. This will ensure that they don't merely trace.

• In the blank space (problem 4) of a *Mathercise* transparency, or on a copy of a master, write an actual problem from a past homework assignment. Or, write a problem of the type found on college entrance exams, or any other type of skill you wish to have your students practice such as estimation or graphing.

• Make available to students, as they enter class, answer strips of paper large enough to work all four problems (create your own or use the answer sheet masters provided with *Mathercise Book D*).

• Immediately following the first bell (signalling the beginning of the passing period between classes), place the *Mathercise* transparency on the overhead projector or pass out individual copies.

• Give students 10-15 minutes after the second bell to work all four problems. Meanwhile, take attendance, return papers, or attend to other classroom chores. You can also use this time to circulate and look at homework.

• Encourage students to work with pencil and straightedge, carefully sketching what they see on the *Mathercise* master.

• Let your students correct their own papers as you demonstrate solutions. Or, ask students to trade papers and correct their neighbors' work. You may find that you need to correct the sketch (problem 3) yourself.

• There are a variety of ways to use *Mathercise*. At the beginning of the year you may wish to use *Mathercise* daily to get students used to the idea of coming to class on time, then move to a once-a-week schedule. Another alternative is to use each *Mathercise* for three days. For example, on Monday your students could work problem 1 (Reason) and problem 4 (Review). On Wednesday, students could work problem 2 (Solve) and a new problem 4. On Friday your students could work problem 3 (Sketch) and another problem 4.

Tips for Doing *Mathercise* Problems

Mathercise problems, while not routine, are self-explanatory enough to not require a lot of teaching. They become gradually more challenging and students learn to do them with practice. Initially, though, you'll want to give students some tips for how to solve different types of problems that they may not have much experience with. What follows are tips for some such problem types found in problems 1 and 3. You can use these examples to teach these problem types when students first encounter them.

Tips for Problem 1 (Reason)

The first problem in each *Mathercise* (Reason) is an inductive or deductive reasoning problem. In inductive reasoning problems the task is fairly straightforward, the student must find the next term in a number or picture pattern. There are, however, a number of different types of deductive reasoning problems. You should work the following examples with your students to familiarize them with these types of problems.

Following Directions

Some students may be unfamiliar with North, South, East, and West. Show students examples of Northwest, Southwest, Northeast, and Southeast. You should also review with them *right, left,* and *about face* turns. A turn to the right or left means a turn of 90°. An about face means a turn of 180°. The best approach to solving this type of problem is to teach students to move their pencils through the steps of the problem (let their fingers do the walking). This problem type is first encountered in Mathercises D5 and D7.

Example

- **Problem:** You are facing west. You turn right, then about face, then left. Which direction is to your right?

- **Solution:**

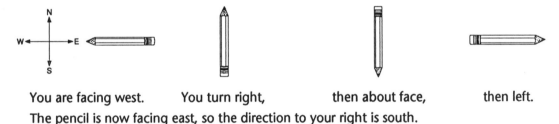

| You are facing west. | You turn right, | then about face, | then left. |

The pencil is now facing east, so the direction to your right is south.

Verbal Reasoning

Many students need to be convinced that it is OK to read and reread a problem over and over again. You should model the reasoning necessary to solve this problem type, reading and rereading the statements slowly and carefully, working your way backwards through the problem. This problem type is first encountered in Mathercises D6 and D8.

Example

- **Problem:** Print the word that is repeated three times in this sentence and underline the vowels in the word.

- **Solution:** The first task is to find a word that is repeated three times in the sentence. The word is *the*. What do we do with this word? Underline the vowels. Therefore, the answer is th<u>e</u>.

Logical Ranking

Drawing diagrams in an organized way helps to solve this type of problem. Model the reasoning necessary to solve this problem type. I recommend that you begin by drawing your translation of the first statement. Then you should draw a new improved ranking based on the second sentence. Continue in this manner until the problem is solved. This problem type is first encountered in Mathercises D15 and D17.

Example

- **Problem:** Annie is funnier than Bonnie. Cara is funnier than Donna. Cara is not as funny as Bonnie. Which one is the funniest? Who is the least funny?

- **Solution:**

ANNIE		ANNIE	?	CARA		ANNIE
BONNIE		BONNIE		DONNA		BONNIE
						CARA
						DONNA

Annie is funnier than Bonnie.	Cara is funnier than Donna.	Cara is not as funny as Bonnie.

Logical Matching

Visual thinking is very helpful in solving this type of problem. To model the reasoning necessary to solve this problem type, I recommend that you begin by making a chart with names on one vertical side and occupations across the top. Then systematically cross out boxes that are logically eliminated and place circles in boxes that are logically matched. This problem type is first encountered in Mathercises D16 and D18.

Example

- **Problem:** Ace, Borus, and Cathy have careers as airplane mechanic, pilot, and air traffic controller but not necessarily in that order. The airplane mechanic works with Ace. Cathy hired the pilot to fly her to Mexico. Borus earns less than the airplane mechanic, but more than the air traffic controller. Match the names and occupations.

- **Solution:**

	Mech	Pilot	Controller
Ace			
Borus			
Cathy			

	M	P	C
A	X		
B			
C		X	

	M	P	C
A	X		
B	X		
C	O	X	X

	M	P	C
A	X	X	O
B	X	O	X
C	O	X	X

Ace, Borus, and Cathy have careers as airplane mechanic, pilot, and air traffic controller.	The airplane mechanic works with Ace. Cathy hired the pilot to fly her to Mexico.	Borus earns less than the airplane mechanic,	but more than the air traffic controller.

Therefore Ace is the air traffic controller, Borus the pilot, and Cathy the mechanic.

Reasonable 'Rithmetic

To model the reasoning necessary to solve this problem type, I recommend that you begin by finding what letter you can first determine. Then rewrite the problem replacing the solved letter with the number. This problem type is first found in Mathercises D25 and D27.

Example

- **Problem:** Each of the letters in the sum on the right represents a different digit. What is the value of A?

$$\begin{array}{r} A\ B \\ +\ B\ A \\ \hline 1\ B\ 5 \end{array}$$

- **Solution:** The first thing to notice is that $A + B$ adds up to 15 and not just 5, because $A + B$ is also in the tens column. Therefore, in the tens column, $A + B +$ (the carry of 1) gives a total of 16. Thus $B = 6$ and $A = 9$.

Bagels

Bagels began as a logical guessing game on computers. In a bagels puzzle you are to determine a 3-digit number (no digit repeated) by taking "educated guesses." After each guess you are given a clue about your guess. The clues:

bagels: No digit is correct.
pico: One digit is correct but in the wrong position.
fermi: One digit is correct and in the correct position.

In each bagels problem, a number of guesses have been made with the clue for each guess shown to the right. From the given set of clues, you can determine the 3-digit number.

Because of space limitations, the meaning of the clues bagels, pico, and fermi will not appear with the *Mathercise* bagels problems. Therefore, you should make a poster of the rules for bagels and post it for use throughout the year (or at the very least, write it on the chalkboard on the day it appears so your students will have it while they work). To model the reasoning necessary to solve this problem type, I recommend that you begin by making a list of all the digits 0 through 9 and then cross out all those that can be eliminated. Note that if a digit from 0 through 9 has not been eliminated, it can appear in the answer, even if it does not appear in the problem. This problem type is first encountered in Mathercises D26 and D28.

Example

- **Problem:**

```
234  pico
567  pico
891  fermi
641  bagels
825  pico
???
```

- **Solution:**

2 3 4̸	P		2 3 4̸	P		2 3 4̸	P		②3 4̸	P
5 6̸ 7	P		5 6̸ 7	P		5̸ 6̸ 7	P		5̸ 6̸ ⑦	P
8 9 1̸	F	→	8 9 1̸	F		8̸ ⑨1̸	F		8̸ ⑨1̸	F
6̸ 4̸ 1̸	B		6̸ 4̸ 1̸	B		6̸ 4̸ 1̸	B		6̸ 4̸ 1̸	B
8 2 5	P	→	8 2 5	P		8̸ 2 5̸	P		8̸ ②5̸	P

```
☐ ☐ ☐        ☐ 9 ☐        ☐ 9 ☐        7 9 2
```

The first thing to notice is that 1, 4, and 6 are eliminated (from line 4, 641 = bagels). Therefore, one of the numbers is either a 2 or a 3, but not both. One of the numbers is either a 5 or a 7, but not both. And one of the numbers is either an 8 or 9, but not both.

We can eliminate the 8 because if 8 is correct, then it must be in the first or left hand position (line 3 = fermi). But line 5 has a pico with an 8 in the first or left most position. This is a contradiction. If 8 is eliminated and 1 is eliminated, then 9 must be correct in the center position.

If 8 is eliminated, then in line 5 either 2 or 5 is correct but in the wrong position. 5 can be eliminated because if 5 is correct, then it must be in the center position (line 2 says 5 cannot be in first position and line 5 says 5 cannot be in the third position). But this is a contradiction because 9 must be in the center position.

Therefore 5 is eliminated and 2 must be correct in the third position. If 5 and 6 are eliminated, then 7 is left for the first position and thus our solution is 792.

Tips for Problem 2 (Solve)

The second problem (Solve) in each *Mathercise* is a symbolic thinking problem. By *symbolic thinking* we mean the problem usually requires the use of algebra or the ability to use symbols to solve it. Each problem reviews one of a number of basic mathematical skills (averages, percent, probability, algebra, proportional thinking, functions, and coordinate geometry). Many of these problems can be found in college entrance exams such as the SAT. Your students should be familiar with most of the types of problems found in the problem 2 portion of the *Mathercise*, except perhaps the problems we call *Fantasy Functions*. These problems appear quite often in SAT exams.

Fantasy Functions

Most students will be unfamiliar with the concept of creating new binary operations. You should show your students examples of new fantasy functions and how to evaluate them for given values. In *Mathercise Book D* this problem type is first encountered in Mathercises D10 and D23.

Example

- **Problem:** If $a \triangle b = a^2 - 2b$, and $f \triangle 5 = f + 32$, where $f > 0$, find f.

- **Solution:** The rule: $a \triangle b = a^2 - 2b$, says, whatever is to the left (a) of the binary operator "$\triangle$" gets squared (a^2) and whatever is to the right (b) of the binary operator "$\triangle$" gets doubled ($2b$). For example, if $a = 3$ and $b = 4$, then $3 \triangle 4 = (3)^2 - 2(4)$ or 1. Therefore, $f \triangle 5 = (f)^2 - 2(5) = f^2 - 10$. But it was given that $f \triangle 5 = f + 32$, therefore $f^2 - 10 = f + 32$. Solving for f we get:

$$f^2 - f - 42 = 0$$
$$(f + 6)(f - 7) = 0$$
Therefore, $f = {}^-6$ or 7
But $f > 0$, therefore $f = 7$.

Tips for Problem 3 (Sketch)

The third problem in each *Mathercise* (Sketch) is a visual thinking problem that requires students to think in two and three dimensions and be able to graph linear, absolute value, quadratic, and exponential functions — many with horizontal and/or vertical shifts. You should practice horizontal and vertical shifts of these functions with your students.

Answers for *Mathercise Book D*

Mathercise D1	1. 49	2. $\frac{3}{100}$ or .03	3. 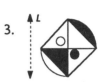
Mathercise D2	1.	2. 78	3.
Mathercise D3	1. 11	2. 8 lbs.	3. A, C, E
Mathercise D4	1. 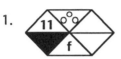	2. 9 cm	
Mathercise D5	1. South	2. $c = 24$	
Mathercise D6	1. t	2. $s = 70$	3.
Mathercise D7	1. Southwest	2. 70	3.
Mathercise D8	1. r	2. 8	3.
Mathercise D9	1. Northwest	2. $2x^2 + 3$	3.
Mathercise D10	1. *SDRAWKCAB*	2. 12	3.
Mathercise D11	1. 81	2. 13	3.
Mathercise D12	1. 	2. $y = 3x - 3$	3. 6 cm
Mathercise D13	1. 77	2. $\frac{1}{4}$	
Mathercise D14	1. 381 365 / 4 / 397 381	2. $20	

| Mathercise D15 | 1. Cynthia | 2. 79 | 3.  |

Mathercise D15   1. Cynthia   2. 79   3.

Mathercise D16   1. Andy – Bricklayer   2. 2.57   3.
Bernice – Contractor
Cassie – Architect

Mathercise D17   1. Jason   2. 120 m   3.
Anne Marie
Darren
Christopher

Mathercise D18   1. Cat – Sushi   2. $y = \frac{3}{2}w$ 3. Yes
Dog – Bruce
Fish – Felix
Snake – Argyle

Mathercise D19   1. No, Amy is strongest.   2. $z = \frac{1}{2}x$

Mathercise D20   1. Jones – Score: 1150   2. $y = 4w$

Mathercise D21   1. ‾1   2. 4   3.

Mathercise D22   1.   2. $9x^2 - 6x + 1$ 3.
or $(3x - 1)^2$

Mathercise D23   1. 5   2. ‾10   3.

Mathercise D24   1.   2. $y = \frac{2}{3}x - \frac{17}{3}$ 3.

Mathercise D25   1. $B = 5$ 2. $y = \frac{-5}{3}x + \frac{19}{3}$ 3. See Answer Below

Mathercise D26   1. 720   2. $\frac{1}{12}$ 3. See Answer Below

Mathercise D27   1. $F = 6$ 2. 50%   3. See Answer Below

Mathercise D28   1. 432   2. 12   3. See Answer Below

D25.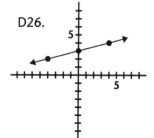

D26.

D27.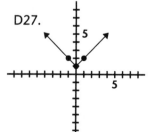

D28.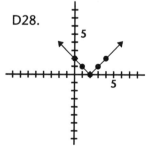

Mathercise D29	1. 4	2. $\frac{3}{8}$ cups water and $4\frac{1}{2}$ cups oats	
Mathercise D30	1. 920	2. $y = \frac{10}{9}w$	

Mathercise D31   1. D   2. $a = 10c$ 3.

Mathercise D32   1.   2. $C = \frac{5}{4}b$ 3. or

Mathercise D33   1. W   2. $x^2 + 3$ 3. No

Mathercise D34   1.   2. 2   3.

Mathercise D35   1.   2. $y = \frac{-3}{2}x + 6$ 3. See Answer Below

Mathercise D36   1. Two   2. $y = \frac{-4}{5}x + 6$ 3. See Answer Below

Mathercise D37   1. One   2. $\frac{1}{9}$ 3. See Answer Below

Mathercise D38   1. North   2. $\frac{5}{36}$ 3. See Answer Below

D35.

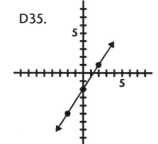

D36.

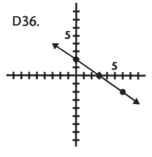

D37.

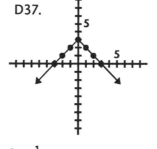

D38.

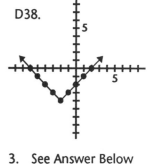

Mathercise D39   1. more   2. $\frac{1}{6}$ 3. See Answer Below

Mathercise D40   1. Carole   2. 86.5   3. See Answer Below

Mathercise D41   1. U   2. $8\frac{1}{3}$ hours   3. See Answer Below

Mathercise D42   1.   2. $y = \frac{3}{2}z$ 3. See Answer Below

D39.

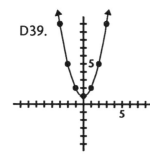

D40.

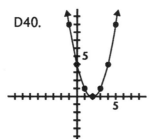

D41.

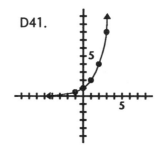

D42.

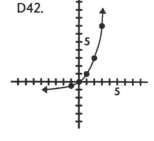

Mathercise D43	1. 69	2. 29	3. See Answer Below
Mathercise D44	1.	2. $c = \frac{1}{d}$	3. See Answer Below
Mathercise D45	1. Cooke, 130	2. $f = {}^-5$	3. See Answer Below
Mathercise D46	1. $A = 4$	2. $k = 18$	3. See Answer Below

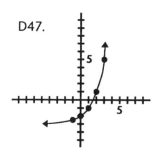

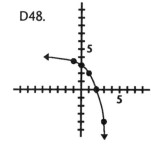

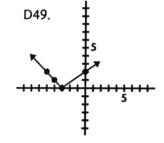

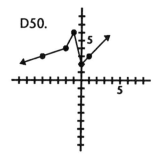

Mathercise D47	1. 781	2. $y = \frac{{}^-7}{2}x + \frac{13}{2}$	3. See Answer Below
Mathercise D48	1. Soto, 190	2. 6	3. See Answer Below
Mathercise D49	1. $C = 8$	2. $\frac{1}{6}$	3. See Answer Below
Mathercise D50	1. 046	2. $\frac{1}{18}$	3. See Answer Below

NAME _____
DATE _____ PERIOD _____
MATHERCISE _____

1.

2.

3.

4.

NAME _____
DATE _____ PERIOD _____
MATHERCISE _____

1.

2.

3.

4.

NAME _____
DATE _____ PERIOD _____
MATHERCISE _____

1.

2.

3.

4.

NAME _____
DATE _____ PERIOD _____
MATHERCISE _____

1.

2.

3.

4.

MATHERCISE D1

1. Reason.

What comes next in the pattern?

13, 24, 17, 29, 21, 34, 25, 39, 29, 44, 33, __?__ ,

2. Solve.

What is 10 % of 20% of $\frac{3}{2}$?

3. Sketch.

Sketch and label right triangle *ABC* with angle *A* the right angle and side *BC* twice the length of side *AC*.

4. Review.

MATHERCISE D2

1. Reason.

What comes next in the pattern?

      ?
___

2. Solve.

If 2 students averaged 72 on a test and 3 others averaged 82, what is the average of all 5 students?

3. Sketch.

Sketch what this figure would look like if it were reflected about the dotted line L.

4. Review.

MATHERCISE D3

1. Reason.

What comes next in the pattern?

0, 1, -2, 3, -4, 5, -6, 7, -8, 9, -10, __?__ ,

2. Solve.

At her last barbecue Char Cole prepared 9 burgers from 2 lbs. of hamburger. She is planning another barbecue and needs 36 burgers. How many pounds of hamburger does she need?

3. Sketch.

If the belt in the figure moves in the direction of the arrow, which of the pulleys move(s) counterclockwise?

4. Review.

©1992 by Michael Serra. Published by Key Curriculum Press, P.O. Box 2304, Berkeley, CA 94702.

MATHERCISE D4

1. Reason.
What comes next in the pattern?

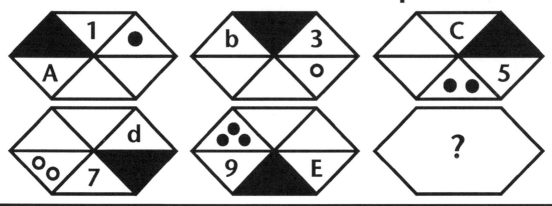

2. Solve.

A mass of 28 kilograms stretches a heavy spring 12 cm. If the distance a string is stretched is proportional to the attached mass, how far will the spring be stretched by a mass of 21 kilograms?

3. Sketch.

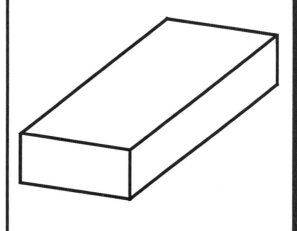

4. Review.

MATHERCISE D5

1. Reason.

You are facing west. You turn right, then about face, then left. Which direction is to your right?

2. Solve.

$$\frac{a}{b} = \frac{c}{d}$$

If $a = 12$, $b = 36$, and $d = 72$, find c.

3. Sketch.

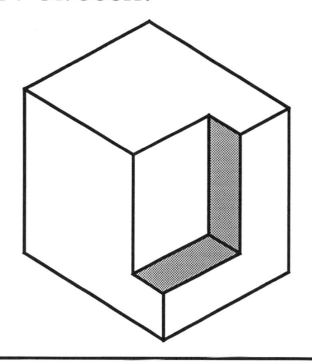

4. Review.

MATHERCISE D6

1. Reason.

In the word *egotistical*, what letter comes after the third letter before the letter *c*?

2. Solve.

$$v = s + \frac{1}{2}at^2$$

If $v = 214$, $a = 32$, and $t = 3$, find s.

3. Sketch.

Sketch this figure resting on its shaded faces.

4. Review.

MATHERCISE D7

1. Reason.

You are facing northwest. You turn right, then right again, then about face, then left. Which direction are you now facing?

2. Solve.

$$K = \frac{n(n-3)}{2}$$

$$P = n^2 + n$$

If $n = 7$, find $P + K$.

3. Sketch.

Sketch what this shape would look like if it were folded along the dotted lines into a solid figure.

4. Review.

MATHERCISE D8

1. Reason.

What is the letter in the word *California* which is two letters after the second consonant that follows the first *a*?

2. Solve.

$f(x) = 2x + 3$
$g(x) = x^2$

Find
$f(-2) + g(-3)$.

3. Sketch.

Sketch one of the possible shapes formed when you place the shaded faces of one solid against the second.

4. Review.

MATHERCISE D9

1. Reason.

You are facing northwest. You turn right, then right again, then right once more, then about face. Which direction is to your left?

2. Solve.

$f(x) = 2x + 3$
$g(x) = x^2$

Find $f(g(x))$.

3. Sketch.

Sketch what this circular figure would look like if it were rolled half its circumference and stopped.

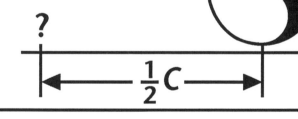

? $\overleftrightarrow{\quad\quad \frac{1}{2}C \quad\quad}$

4. Review.

MATHERCISE D10

1. Reason.

If crossing out the *M, H,* and the third letter of the alphabet in the word *MATCH* leaves a meaningful two–letter word, print the word *BACKWARDS*, backwards. Otherwise print the word *FORWARD* backwards.

2. Solve.

$a * b = 2b - ab$

Find $(3 * 2) * 3$.

3. Sketch.

Sketch and label rhombus *ABCD* with the measure of angle *A* about 120 degrees.

4. Review.

MATHERCISE D11

1. Reason.

What comes next in the pattern?

18, 14, 21, 13, 26, 12, 33, 11,
42, 10, 53, 9, 66, 8, __?__ ,

2. Solve.

What is the length of segment *AB* if *A* has coordinates (0, 5) and *B* has coordinates (12, 10)?

3. Sketch.

Sketch what this figure would look like if it were rotated clockwise 90° about the point *T*.

4. Review.

©1992 by Michael Serra. Published by Key Curriculum Press, P.O. Box 2304, Berkeley, CA 94702.

MATHERCISE D12

1. Reason.

What comes next in the pattern?

	1	
4	5	2
	3	

10		8
	6	
12		10

	24	
28	7	24
	28	

59		55
	8	
63		59

	122	
130	9	122
	130	

?

2. Solve.

What is the equation in slope-intercept form of the line with a slope of 3 passing through (2, 3)?

3. Sketch.

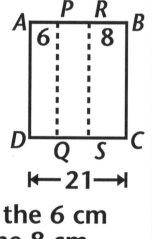

Rectangle *ABCD* represents a sheet of paper 21 cm wide. The dotted lines *PQ* and *RS* are fold lines. When the 6 cm flap *APQD* and the 8 cm flap *RBCS* are folded over on top of the center, what is the width of the overlap?

4. Review.

MATHERCISE D13

1. Reason.

What comes next in the pattern?

7, 21, 35, 49, 63, __?__ ,

2. Solve.

What is the probability of spinning a sum of 5 with the two spinners shown?

3. Sketch.

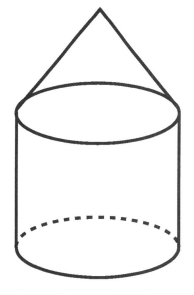

4. Review.

MATHERCISE D14

1. Reason.

What comes next in the pattern?

```
          1
    7     9     3
          5
```

```
   17          13
          8
   21          17
```

```
          38
    46    7     38
          46
```

```
   91          83
          6
   99          91
```

```
          180
   196    5    180
          196
```

```
          ?
```

2. Solve.

How much interest is earned in 4 months on $1000 deposited at 6% annual interest compounded 3 times a year?

3. Sketch.

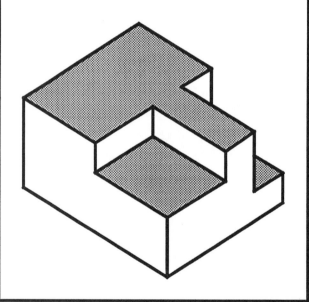

4. Review.

MATHERCISE D15

1. Reason.

Art, Bozo, Cynthia, and Doris are stand-up comics. Art is not as funny as Bozo but funnier than Doris. Cynthia is not as funny as Bozo but funnier than Art. Who is funnier, Cynthia or Doris?

2. Solve.

If Igor's first 2 test scores are 100 and 91, what is the minimum score that he must get on the third test to get an average of at least 90?

3. Sketch.

Sketch this figure resting on its shaded faces.

4. Review.

MATHERCISE D16

1. Reason.

Andy, Bernice, and Cassy have jobs as architect, bricklayer and contractor but not necessarily in that order. The contractor does consulting work for Andy's business. Bernice earns less than the architect but more than Andy. Match names with occupations.

2. Solve.

A pitcher allowed 12 earned runs in 42 innings. What is the pitcher's earned run average? That is, how many runs would be allowed in 9 innings? Round your answer to the nearest 100th.

3. Sketch.

Sketch what this shape would look like if it were folded along the dotted lines into a solid figure.

4. Review.

MATHERCISE D17

1. Reason.

Ann Marie scored 30 more points on her achievement test than Christine but 20 less than Jason. Jason has 30 more than Darren. Rank them from the student with the highest achievement score on top to the student with the lowest on the bottom.

2. Solve.

A wire of uniform thickness and density weighing 96 kilograms is cut into 2 pieces. One piece is 90 meters long and weighs 72 kilograms. What is the length in meters of the original?

3. Sketch.

Sketch one of the possible shapes formed when you place the shaded faces of one solid against the second.

4. Review.

MATHERCISE D18

1. Reason.

Igor has a cat, small dog, large goldfish, and a snake. Their names are Argyle, Bruce, Sushi, and Felix. Sushi is the biggest, the snake is the smallest, the cat is younger than Bruce but older than Argyle, and the goldfish is bigger than Argyle but smaller than Bruce. Felix is not the cat. Match each pet with its name.

2. Solve.

$$y = 2x$$
$$x = \frac{3}{4}w$$

Find y in terms of w.

3. Sketch.

Are the two shapes that make up the rectangle identical? (Ignoring the shading!)

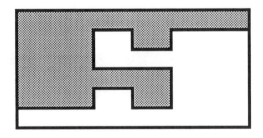

4. Review.

MATHERCISE D19

1. Reason.

Amy is stronger than Bonnie. Carmen is stronger than Dina. Carmen is weaker than Bonnie. Eleanor is stronger than Carmen but not as strong as Amy. From the information given, is it possible to determine whether Eleanor is stronger than Bonnie? Which one is the strongest of the five?

2. Solve.

$x = 3y$

$y = \frac{w}{6}$

$w = 4z$

Find z in terms of x.

3. Sketch.

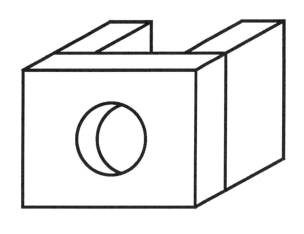

4. Review.

MATHERCISE D20

1. Reason.

Huey, Duey, and Louie differ in SAT scores. Their last names are Swift, Jones and Black, but not necessarily in that order. Huey's SAT score is 150 points higher than Louie's but 70 points lower than Duey's. Swift had the highest of the three scores (1220) and Black the lowest. What is Huey's last name and what was his score?

2. Solve.

$$3x = 2y$$
$$3x = 8w$$

Find y in terms of w.

3. Sketch.

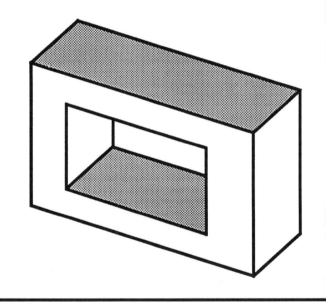

4. Review.

MATHERCISE D21

1. Reason.

What comes next in the pattern?

5, 6, 4, 7, 3, 8, 2, 9, 1, 10, 0, 11, ___?___ ,

2. Solve.

$f(x) = 3x - 1$
$g(x) = x^2 + 1$

Find
$g(f(0)) + f(g(0))$.

3. Sketch.

Sketch one of the possible shapes formed when you place the shaded faces of one solid against the second.

4. Review.

MATHERCISE D22

1. Reason.

What comes next in the pattern?

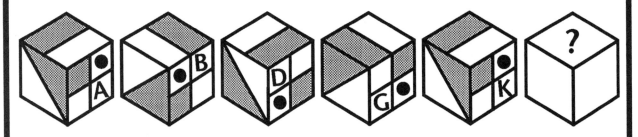

2. Solve.

$f(x) = 3x - 1$
$g(x) = x^2$

Find $g(f(x))$.

3. Sketch.

Sketch what this shape would look like if it were folded along the dotted lines into a solid figure.

4. Review.

MATHERCISE D23

1. Reason.

What comes next in the pattern?

1, 0, 2, 0, A, 3, 0, A, B, 4, 0, A, B, C, __?__ ,

2. Solve.

$a \circ b = ab - 2(a + b)$

Find $3 \circ (2 \circ 3)$.

3. Sketch.

Sketch one of the possible shapes formed when you place the shaded faces of one solid against the second.

4. Review.

MATHERCISE D24

1. Reason.

What comes next in the pattern?

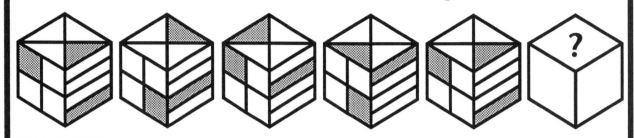

2. Solve.

What is the equation in slope-intercept form of the line with a slope of $\frac{2}{3}$ passing through (1, -5)?

3. Sketch.

Sketch one of the possible shapes formed when you place the shaded faces of one solid against the second.

4. Review.

MATHERCISE D25

1. Reason.

Each of the three letters in the sum on the right represents a different digit. What is the value of *B*?

$$
\begin{array}{r}
3\ 7\ 2 \\
3\ 8\ 4 \\
+\ 9\ B\ 4 \\
\hline
C\ 7\ C\ A
\end{array}
$$

2. Solve.

What is the equation in slope-intercept form of the line passing through (2, 3) and (5, -2)?

3. Sketch.

Sketch the graph of $y = \frac{3}{2}x + 1$.

4. Review.

©1992 by Michael Serra. Published by Key Curriculum Press, P.O. Box 2304, Berkeley, CA 94702.

MATHERCISE D26

1. Reason.

From the given set of clues, find a 3-digit number that satisfies all the clues.

```
123  fermi
567  pico
340  fermi
150  fermi
146  bagels
???
```

2. Solve.

What is the probability of rolling a sum of 4 with a pair of 6-sided dice?

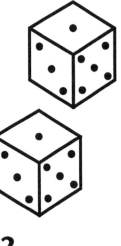

3. Sketch.

Sketch the graph of the line through $(0, 3)$ with a slope of $\frac{1}{4}$.

4. Review.

MATHERCISE D27

1. Reason.

Each of the six letters in the product on the right represents a different digit. What is the value of *F*?

```
      B  C  6
   X     A  7
  ─────────────
   A  A  C  A
   F  B  A
  ─────────────
   E  D  B  A
```

2. Solve.

If 70% of *A* is equal to 35% of *B*, then *A* is what percent of *B*?

3. Sketch.

Sketch the graph of
$y = |x| + 1$

4. Review.

MATHERCISE D28

1. Reason.

From the given set of clues, find a 3-digit number that satisfies all the clues.

```
729  pico
367  pico
480  fermi
687  bagels
493  fermi pico
???
```

2. Solve.

The average of 3 numbers is 5 and the average of 7 other numbers is 15. What is the average of all 10 numbers?

3. Sketch.

Sketch the graph of $y = |x - 2|$

4. Review.

MATHERCISE D29

1. Reason.

Each of the 3 letters in the sum on the right represents a different digit. What is the value of C?

```
   8 7 8 9
   3 B A 7
   4 8 2 A
 + 7 A B 5
 ─────────
 2 C 2 8 7
```

2. Solve.

An oatmeal cookie recipe makes 4 dozen cookies and calls for a $\frac{1}{4}$ cup of water and 3 cups of oats. To make 6 dozen cookies how much water and oats are necessary?

3. Sketch.

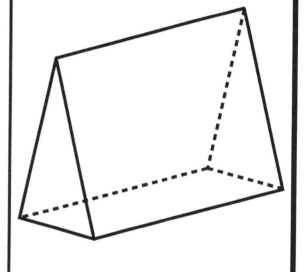

4. Review.

MATHERCISE D30

1. Reason.

From the given set of clues, find a 3-digit number that satisfies all the clues.

```
789  pico
567  bagels
480  fermi
390  pico  fermi
493  pico
910  fermi  fermi
???
```

2. Solve.

$$2x = 3y$$
$$3x = 5w$$

Find y in terms of w.

3. Sketch.

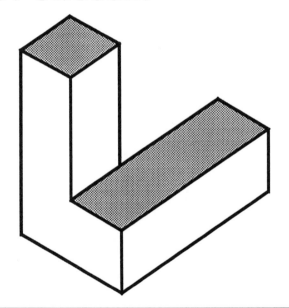

4. Review.

MATHERCISE D31

1. Reason.

What comes next in the pattern?

5, A, 10, Z, 15, B, 20, Y, 25, C, 30, X, 35, __?__ ,

2. Solve.

$a = 4b$
$2b = 5c$

Find a in terms of c.

3. Sketch.

Sketch this figure resting on its shaded faces.

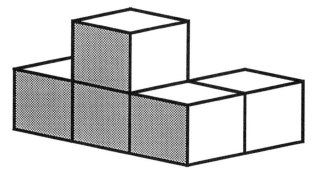

4. Review.

MATHERCISE D32

1. Reason.

What comes next in the pattern?

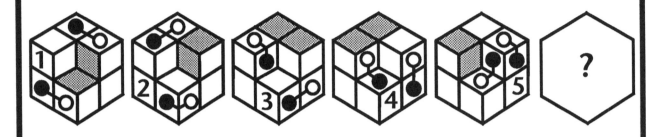

2. Solve.

$$3A = 5B$$
$$6A = 8C$$

Find C in terms of B.

3. Sketch.

Sketch what this shape would look like if it were folded along the dotted lines into a solid figure.

4. Review.

MATHERCISE D33

1. Reason.

What comes next in the pattern?

Z, 1, A, 22, Y, 333, B, 4444, X, 55555, C, 666666, ___?___ ,

2. Solve.

$$f(x) = 4x^2 + 3$$
$$g(x) = \frac{1}{2}x$$

Find $f(g(x))$.

3. Sketch.

Are the two shapes that make up the rectangle identical?
(Ignoring the shading!)

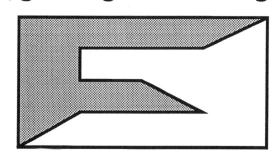

4. Review.

MATHERCISE D34

1. Reason.

What comes next in the pattern?

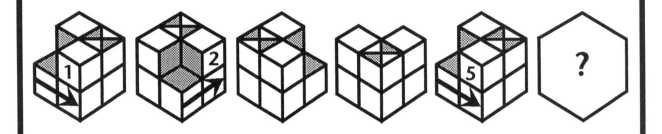

2. Solve.

$a \triangle b = a^b$

Find
$2 \triangle (5 \triangle 0)$.

3. Sketch.

Each pair of opposite faces on the cube below are identical except that the shading is reversed. Sketch a view of the cube showing the three hidden faces.

4. Review.

MATHERCISE D35

1. Reason.

There are exactly 5 houses on a block running North and South. The green and red houses are to the north of the yellow house. The yellow house is adjacent to the blue and orange houses. The red house is between the orange and green houses. Draw 5 squares in a vertical column representing the five houses and label them according to color and label North-South.

2. Solve.

What is the equation in slope-intercept form of the line passing through (2, 3) and parallel to the line $y = \frac{-3}{2}x - 2$?

3. Sketch.

Sketch the graph of the line with y-intercept (0, -2) and parallel to $y = \frac{3}{2}x + 1$.

4. Review.

©1992 by Michael Serra. Published by Key Curriculum Press, P.O. Box 2304, Berkeley, CA 94702.

MATHERCISE D36

1. Reason.

The letter in the word *melancholy* which is two letters before the third vowel appears in how many different words in this problem?

2. Solve.

What is the equation in slope-intercept form of the line passing through (0, 6) and perpendicular to the line $y = \frac{5}{4}x + 1$?

3. Sketch.

Sketch the graph of the line with x-intercept (3, 0) and perpendicular to $y = \frac{3}{2}x + 1$.

4. Review.

MATHERCISE D37

1. Reason.

At the company picnic Patrick ate 1 more helping of salad than Mike but 3 fewer helpings of sushi. Mike ate 2 more helpings of salad than Ingrid but 2 fewer helpings of sushi than Ingrid. If Ingrid ate 6 helpings of sushi and 3 helpings of salad how many helpings of sushi did Patrick eat?

2. Solve.

What is the probability of rolling a sum of 5 with a pair of 6-sided dice?

3. Sketch.

Sketch the graph of the absolute value function
$$y = -|x| + 3$$

4. Review.

MATHERCISE D38

1. Reason.

I am facing the front of my house. While looking out my front window I see a horse pass by the front of the house walking toward the setting sun with my house at his left. Which direction am I facing?

2. Solve.

What is the probability of rolling a sum of 6 with a pair of 6-sided dice?

3. Sketch.

Sketch the graph of the absolute value function $y = |x + 2| - 4$

4. Review.

MATHERCISE D39

1. Reason.

What is the first word in this problem that starts with a consonant and ends with a vowel and has more than three letters?

2. Solve.

A mixture consists of .25 kg of pecans and 1.25 kg of walnuts. What fraction of the mix, by weight, is pecans?

3. Sketch.

Sketch the graph of $y = x^2 + 1$

4. Review.

MATHERCISE D40

1. Reason.

Aaron is shorter than Bernice or Carole, but is taller than Dede and Ernesto. Ernesto is the shortest and lightest, shorter than the second shortest by 2 inches and lighter than the second lightest, Aaron, by 8 pounds. Carole is lighter than Bernice but heavier than Dede. None of the 5 is both the heaviest and tallest. Who is tallest?

2. Solve.

A class of 32 students took a test that was scored from 0 to 100. Eighteen students received scores of 76. What is the highest class average?

3. Sketch.

Sketch the graph of $y = (x - 2)^2$

4. Review.

©1992 by Michael Serra. Published by Key Curriculum Press, P.O. Box 2304, Berkeley, CA 94702.

MATHERCISE D41

1. Reason.

What comes next in the pattern?

1, Z, 4, Y, 9, X, 16, W, 25, V, 36, ___?___ ,

2. Solve.

In the first semester physics course, a student is expected to spend 2 hours studying for every hour of class time. If the class meets 5 days a week for 50 minutes, how many hours of study a week is expected of the students?

3. Sketch.

Sketch the graph of $y = 2^x$

4. Review.

©1992 by Michael Serra. Published by Key Curriculum Press, P.O. Box 2304, Berkeley, CA 94702.

MATHERCISE D42

1. Reason.

If the pattern of figures continues on the dots, what figure will appear on the last dot?

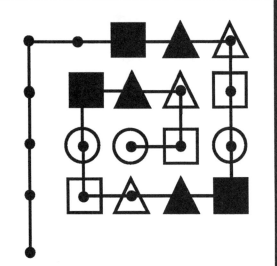

2. Solve.

$y = 3x$
$x = 2w$
$z = 4w$

Find y in terms of z.

3. Sketch.

Sketch the graph of $y = 2^x - 1$

4. Review.

MATHERCISE D43

1. Reason.

What comes next in the pattern?

A, 6, b, 9, C, 15, d, 24, E, 36, f, 51, G, ___?___ ,

2. Solve.

$2x + y = 12$
$3x - 2y = 17$

Find $5x - y$.

3. Sketch.

If the graph is $y = f(x)$, sketch the graph of $y = f(x) + 2$.

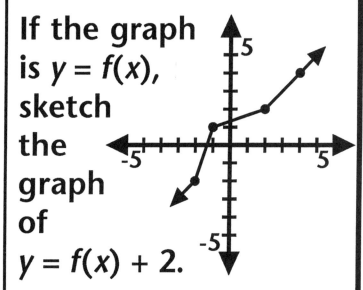

4. Review.

MATHERCISE D44

1. Reason.

If the pattern of figures continues on the dots, what figure will appear on the last dot?

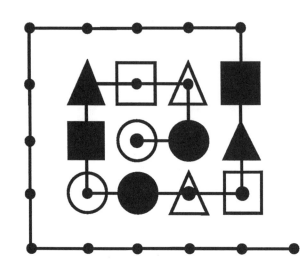

2. Solve.

$c = \frac{e}{f}$

$e = \frac{f}{d}$

Find c in terms of d.

3. Sketch.

If the graph is $y = f(x)$, sketch the graph of $y = f(x + 3)$.

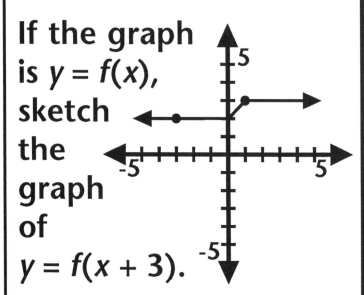

4. Review.

©1992 by Michael Serra. Published by Key Curriculum Press, P.O. Box 2304, Berkeley, CA 94702.

MATHERCISE D45

1. Reason.

Able, Baker, and Charlie differ in their bowling averages. Their last names are Elba, Cooke, and Daniels, but not necessarily in that order. Able's bowling average is 20 pins higher than Charlie's, but 11 pins lower than Baker's. Elba had the highest of the three scores and Daniel's the lowest (110). What is Able's last name and his score?

2. Solve.

$$f(x) = 2x + 3$$
$$g(x) = \frac{1}{2}(x - 3)$$

Find $f(g(-5))$.

3. Sketch.

Sketch the graph of $y = -x^2 - 3$

4. Review.

MATHERCISE D46

1. Reason.

Each of the three letters in the problem on the right represents a different digit. What is the value of *A*?

$$
\begin{array}{r}
A\ B\ C\ C \\
-\ B\ 2\ C\ 6 \\
\hline
1\ 9\ 9\ 7
\end{array}
$$

2. Solve.

$a \# b = ab + a - b$

$6 \# k = 96$

Find *k*.

3. Sketch.

Sketch the graph of
$y = (x + 2)^2 - 4$

4. Review.

MATHERCISE D47

1. Reason.

From the given set of clues, find a 3-digit number that satisfies all the clues.

```
129  pico
567  pico
480  fermi
690  bagels
415  pico
???
```

2. Solve.

What is the equation in slope-intercept form of the line passing through (1, 3) and parallel to the line that passes through the points (0, 1) and (2, -6)?

3. Sketch.

Sketch the graph of $y = 2^x - 3$

4. Review.

MATHERCISE D48

1. Reason.

Adam, Boris, and Coreen differ in their batting averages. Their last names are Smith, Scarpelli, and Soto, but not necessarily in that order. Adam's batting average is 40 points higher than Coreen's but 25 points lower than Boris'. Smith had the highest of the 3 averages and Soto the lowest (190). What is Coreen's last name and her average?

2. Solve.

What is the length of segment MN if M is the midpoint of side AC and N is the midpoint of side BC in triangle ABC with vertices A at $(0, 0)$, B at $(12, 0)$, and C at $(8, 6)$?

3. Sketch.

Sketch the graph of $y = -2^x + 4$

4. Review.

MATHERCISE D49

1. Reason.

Each of the five letters
in the problem on
the right represents
a different digit.
What is the value of C?

```
            5 2
   A  E) C  B  2
        C  D
           B  2
           B  2
```

2. Solve.

What is the
probability
of rolling
a sum
of 7
with a
pair of
6-sided dice?

3. Sketch.

If the
graph is
$y = f(x)$,
sketch
the
graph
of
$y = f(x + 3)$.

4. Review.

©1992 by Michael Serra. Published by Key Curriculum Press, P.O. Box 2304, Berkeley, CA 94702.

MATHERCISE D50

1. Reason.

From the given set of clues, find a 3-digit number that satisfies all the clues.

```
509  pico
867  pico
620  pico  pico
123  bagels
???
```

2. Solve.

What is the probability of rolling a sum of 2 or a sum of 12 with a pair of 6-sided dice?

3. Sketch.

If the graph is $y = f(x)$, sketch the graph of $y = f(x + 3) + 1$.

4. Review.